EDIBLE GARDENING IN THE TROPICS

BY KARIN FIELDS

trimarkpress

Published by TriMark Press, Inc., Deerfield Beach, Florida.

Library of Congress Cataloging-in-Publication Data

Edible Gardening in the Tropics
Karin fields
Cover image by Leesa Richards

p. cm.

ISBN: 978-1-943401-52-9

Library of Congress Control Number: 2018961828

K18

10 9 8 7 6 5 4 3 2 1

First Edition

Printed and Bound in the United States of America

publishing the written word
trimarkpress

368 South Military Trail
Deerfield Beach, FL 33442
800.889.0693
www.TriMarkPress.com

Karin speaking to the Master Gardeners at the Institute of Food & Agricultural Sciences in 2017.

DEDICATION

To all the Wards of the State ,
my bothers and my sisters:

"What the mind can conceive and believe,
it can achieve."
-Napoleon Hill

SPECIAL THANKS

To my best bud…my husband!
I am one lucky Edible Gardening Gal

TABLE OF CONTENTS

INTRODUCTION

Kids call vegetables "yucky" and you know what? They're right.

Not only are many vegetables yucky, they are often depleted of much of their nutritional value as well. If you buy a bunch of parsley or a head of lettuce in a grocery store, that produce may have spent at least a week being trucked across the country, stored for days in food distribution centers, and displayed in the store before you buy it. During that week, both the nutritional value and taste of the vegetables have declined. Vitamin C is especially likely to decrease during handling, but other nutrients such as B vitamins are affected as well.

According to an article written by Diane M. Barrett, director for the Center for Excellence in Fruit and Vegetable Quality at the University of California, Davis, if you buy peas that have been stored for one or two days, vitamin C levels will have declined by as much as 61% after home cooking. Other studies have shown similar results for spinach, carrots, broccoli, and green beans.

Buying organic food from a health food store does not solve the problem of nutrient depletion. The news show 60 Minutes reported that Whole Foods buys many of its vegetables from China, which means that the food has been shipped overseas and lost many of its nutrients during that period.

Not only are we Americans eating vegetables that have been stored for too long, we are also eating fewer of them than our ancestors did.

As food journalist and professor at the University of California, Berkeley, Michael Pollan explains that modern humans have shifted from eating leaves to eating seeds like cereal grains.[1] While many factors have led to this dietary change, one of them is cost. Pollan made this point when he spoke at a farm-bill forum in Washington, D.C. Holding up a box of Twinkies and a bunch of carrots, he told the assembled lawmakers that the junk food costs 99 cents while the carrots cost $2.99.

And the vegetable farmers are not getting rich. Compared to grains, vegetables are more labor-intensive. For example, broccoli takes 90 days to grow – 90 days of planting, watering, dealing with pests and diseases, and harvesting each head. The farmer's reward for months of work is extremely slim. This is the reason why our farmers in America are subsidized by the government.

One solution to work around this rapid decline in nutrition after harvest is to grow your own. However anyone who has tried to grow anything knows that gardening involves far more than dropping seeds in the ground and watering them. While you don't need a college degree in agriculture – our grandparents didn't, and they grew enough food to feed a nation – you do need some basic knowledge that is specific to our local climate.

The important word you're looking for when buying food is not "organic." It's LOCAL. By supporting our local farmers who work very hard, you will not only be getting food that is at its optimum nutritional level, but investing in our local economy as well.

You can supplement what you are not getting from conventional food, by buying direct from your local farmers, growing your own and eating frozen. Frozen food is flash frozen right in the field, retaining a lot of its nutrients than so-called "fresh."

When you understand the logic about why you do what you do, you will be better at making decisions for yourself. That's what this book is for. I am going to help you produce great food in a sustainable way by keeping it simple, debunking some gardening myths, and telling it just like it is. My goal is to save you time, money, and frustration while helping you to, in the words of George W. Bush, "put food on your family."

Here's the logic: Edible plants don't necessarily want to grow here in this extreme heat and humidity. The same warm climate that attracts "snow birds" provides a good home for our garden pests as well. The sand that feels so good to lie on at the beach is a major component of our soil and is lacking in nutrients, and does not hold onto water. The humidity in the air can lead to rot and diseases, and plants find the August heat as hard to bear as we do.

Growing in the winter is not a perfect solution either. Shorter days mean that plants get less of the sunlight they need to thrive.

We can overcome these challenges by using this book as a guide.

Commitment is key. If you play a sport or a musical instrument, you know it requires daily effort to get really good. Vegetable gardening is similar. Before planting a garden, ask yourself if you are ready to spend part of each day tending it. Plants are like babies – both require daily work. They need food, water, and good environments that protect them. Just as a figure skater can't neglect practice for weeks and hope to win an Olympic medal, it's not a good idea to neglect your edible plants, and hope to get a major harvest.

Gardening, like any other skill, requires learning. I am self-taught and have learned about growing food mostly through personal experience, watching gardening shows, reading books, and studying privately with my gardening mentors. I am still learning all the time, and most of it is by doing.

I was an organic gardener for many years but have learned through my research that that the word "organic" really is a misnomer. Even organically grown produce contains residues of synthetic pesticides and the "natural" substances are not always free of problems, either. Much of the information I learned about organic gardening didn't make logical sense. I looked at all my plastic bottles of potions for each and every little problem, and changed my mind. I wasn't getting much bounty either. I am now a natural or a more science-based gardener. I check my information now, to make sure it's not gardening woo woo.

After you've learned these basic skills for the tropics, you can decide for yourself what suits you and your life...but you need to walk before you can run. We all have our ways of doing things. Find your way.

I'm greener now by using my own compost, and keeping pesticides to a very bare minimum…if at all.

We are going to start you with commercially prepared vegetable soil mixes because they are the easiest and most effective way of growing vegetables in South Florida. These mixes are specifically formulated and will help you, especially on your first try.

This is so important…success on your first try! Now you will be so interested that you will learn just because. Who wants to read a huge gardening instructional manual…it's really not necessary!

Maybe later…not at first…it could scare you!!!

Start small. Only grow what you really like to eat. You can always expand your garden later.

Oh, and one more thing… the science changed in the late 1980s, so, if you are reading any gardening books from that time period, take the information with a grain of salt!

We sowed, we reaped, and now we eat!

OKAY – LET'S GET STARTED!!!!!!

1. Pollan, Michael. "Unhappy Meals." https://michaelpollan.com/articles-archive/unhappy-meals.

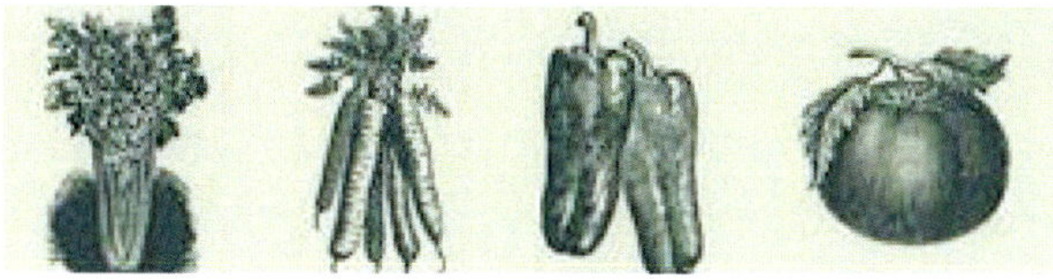

1. MAKE A GARDEN PLAN

Grow Zones

In the 1960s, the USDA divided North America into 10 different Grow Zones. These grow zones indicate and guide you as to what grows best in each particular area of the country. South Floridians live in Zone 10b -11.

A list of the varieties that grow best in our Zone is on page 45.

Think about zoning laws, homeowners' association regulations, and your neighbors. Some gardeners have had to uproot vegetable plots because they unintentionally violated community restrictions. Be cautious about growing vegetables in the front yard...your neighbors may not appreciate this...unless of course you give them some!

Begin by choosing a location for your garden that is close to a water spigot so you'll never have to lug that heavy jug! Be sure to turn off any sprinkler heads in the area that could accidently spray water into your garden or onto your leaves, which could lead to the development of foliar (leaf) diseases.

Make sure the location will provide your edible plants with full sun during the winter months, but that doesn't mean you can't grow in the shade.

Many plants are shade tolerant, meaning, they will grow in part sun/part shade or under a tree. I included a list of shade tolerant plants on page 71. Herbs are very shade tolerant.

If you are using containers, it will be easier to mow around them if you put some landscape fabric under them. Set your containers on top of the fabric.

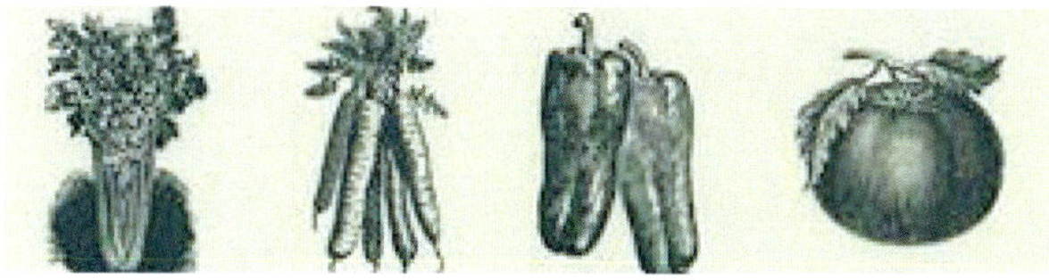

2. ALL ABOUT SEEDS

Starting from seed can give you the most control over what you would like to grow. You can purchase native seeds from Anne in Hallandale, FL at www.eonseed.com. Anne has all the correct seeds for our grow zone.

In addition, you may also try www.johnnyseeds.com, but make sure the variety you pick will grow in Zone 10b. The zone information is contained in the back of the seed pack.

Seeds must remain damp at all times to germinate. If you let them dry out at all during the germination process they will not come back. Different seeds have different shelf lives, so if your germination rate is poor, and you've been keeping them moist, your seeds could be bad. Seeds are cheap so invest in fresh seed…especially for newbies!

September is a good time to start your seeds here. You can attempt to start them earlier, but some seeds will not germinate in the extreme heat.

Seed types:

Heirloom – seed that is produced by open pollination and natural means.

Hybrid – seed that is produced by cross-pollinating. Certain characteristics are bred into one seed to produce a better plant.

GMO – seed that had its DNA altered using techniques of genetic engineering. These seeds are only available to farmers.

Some people soak their seeds in water overnight to soften the seed coat to speed up the germination process. It's not a good idea for the seeds to sit in water because it can create fungal problems.

For a new gardener, using a seed–starting mix will give you the best success for seed germination, and will teach you about sprouting your own seeds.

To start your seeds, fill your cells to the top with your seed–starting mix.

Fill the under-tray with water about 1/4 inch up from the bottom, and let the cells sit in the water until they are completely saturated…maybe 5 minutes. You will notice the soil change to a darker color. Poke a hole about halfway down into the center of each cell. Place one seed in each cell and cover it with your seed-starting mix. Tamp the soil down lightly.

Seed-starting tray cells

Seed starting under-tray

When the cells are completely saturated, remove them from the water, and let the excess drain out the bottom.

As a general rule, the depth at which you plant a seed is about three times the size of the seed, but a seed-starting mix is so light, it doesn't matter all that much about the depth at which you plant it. They'll find their way to the light.

Use the under-tray to water your seedlings. Set the cells in your under-tray for about 5 minutes. The seedlings will take up the water they need.

Misting in the tropics isn't a good idea, period. We have enough mist!

Label each cell with your paint pen. Paint pens last the longest in the sun, and can be bought at any hobby shop. On that label, be sure to include the name, the date, and the variety of seed.

Place your cells in the shade until the first sprout pops up out of the soil, and then place the entire tray in the sun. If you are not going to be home all day, this is where you will utilize that under-tray, being careful to put only a bit of water at the bottom. You don't want them sitting in water all day.

For replanting, the seedlings need to be about two to four weeks old, and have at least two sets of leaves on their stem. They are now ready to be replanted into their permanent homes.

Don't pull the plants out by their stem. Nudge them gently out of their cells/container by gently squeezing the bottom of the cell with one hand, and pouring the entire plant into your other hand. Be careful…it's a baby!

Root crops, like carrots, parsnips, and radishes, don't like to be transplanted. A root crop is a vegetable that grows under the soil line, while the leaves grow above it. The procedure or terminology for planting root crops is called "direct sowing." You will direct sow these seeds into their permanent homes, instead of using a seed tray.

When planting seeds or starter plants, space them evenly apart. Simply envision the size of the vegetable and give it a little more space to allow for its leaf span. For example, a carrot can get to be maybe 3/4 of an inch in diameter, so plant your carrot seeds with that in mind and space those seeds about 2-3 inches apart. How many plants or seeds go into a square foot is discussed under each individual crop listed starting on page 45.

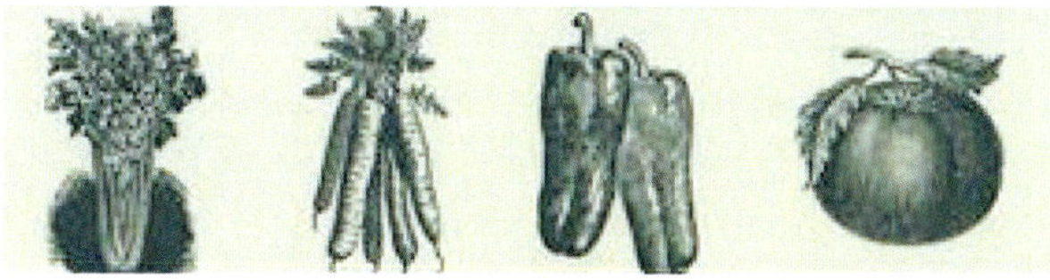

3. STARTER PLANTS

Starter plants can give you a jump on the season, but some of these plants could possibly be infected with bugs or disease. Others could have been started in a greenhouse, and may not yet be acclimated to our weather in the south. If you do use starter plants, make sure they look healthy. Check over and especially **UNDER** the leaves before you buy, to look for signs of pests or disease such as webbing, eggs, spots, or yellowing.

Some stock plants sold in chain stores are inappropriate for our season and area. For example, they may sell tomato plants during the summer when the rest of the country is planting tomatoes, but it's the wrong time of the year to be planting tomatoes in the tropics. You have a better chance of getting plants appropriate for our climate and season if you buy from a reputable nursery, or know your farmer.

When replanting a starter plant into its permanent location look at its roots. If they are circling, carefully break them up and untangle them. We want the roots to be able to go out into the soil like the spokes on a wheel.

Dig a hole twice as wide as the plant, and no deeper than its root ball. Stabilize your plant by hilling around it.

Root-bound starter plant

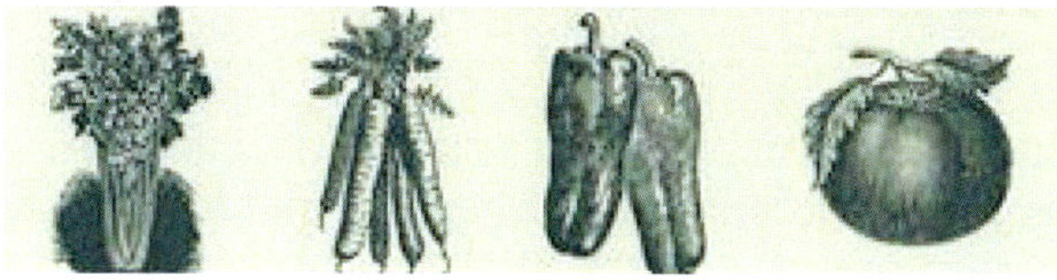

4. CONTAINERS

Our soil is sandy, doesn't hold water, and is not nutritionally sufficient to grow edibles in the tropics. We grow our veggies in containers to contain our "MIX" because it has everything it needs in it, and we've paid good money for it! Mix includes perlite which is puffed mica and makes for better drainage, and less compaction of the soil.

You can do a container in many different ways. Go on the Internet to see what many imaginative people have created!

Avoid any metal or clay containers that will cook your plants. Make sure there is sufficient drainage at the bottom of whatever container you choose. Most of the glazed fancy containers don't have drainage holes. In that case, you will keep your plant in its original container with the good drainage, and place it into that glazed pot. Elevate the inside container by placing several inches of stones at the bottom of the glazed pot so that when it rains, your plant won't drown.

If you decide to build a container/frame, make it no wider than 4 feet across at any one point, to make it easier to reach the middle, without having to step on the soil.

Pine lasts about 2-3 years. Cedar lasts about 12 years. The corners fall apart first so secure your corners well. You can buy the corners instead of using nails. Place professional grade landscape fabric inside the frame and have it drape at least halfway up the sides to keep the nematodes, a microscopic garden pest found in sandy soil, out of our garden. Because the fabric is porous, it will also allow the water to drain out underneath the bed.

Composite lasts forever.

Beds made out of coral, above, and concrete block, below, last forever.

Grow Bags

Grow bags are porous fabric containers that take much of the guesswork out of watering, by allowing the excess water to drain out, thereby eliminating over-watering. They come in many different sizes and are cheap to use – about $10 for a 10-gallon bag. Grow bags can be placed on concrete or grass. They can be used on a terrace with a drainage bowl underneath. Bags are lightweight and moveable. Confuse your pests by moving the bag, making it more difficult for them to find yesterday's dinner! I have many of these bags and love them! If you grow tomatoes in bags, place a plastic bag underneath so the nematodes won't get in…they love tomatoes!

Grow Bag Farm

Self-Watering Containers

These plastic containers have a water reservoir built into the bottom of the box that is capable of irrigating the plants for several days and up to a week in some cases. The EarthBox, the very first self-watering container, was very innovative many years ago and has since been copied. The copies are cheaper and just as good. They can be made at home as well. Look on the Internet for some ideas. This is a great invention, especially for growing tomaties, and worth the extra cost for newbies.

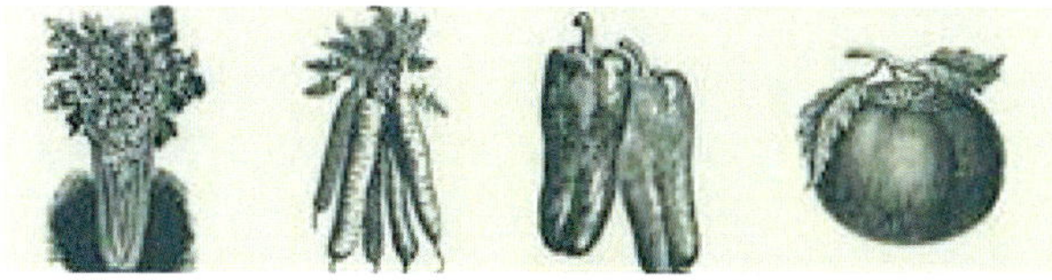

5. SOIL

Soil is alive and loaded with microorganisms that create communities by building homes, paths, and tunnels… just like people and animals. We want a greater variety of microorganisms and biodiversity in our soil because these living, breathing organisms break down the organic matter that nourishes our plants. They also help plants ward off various diseases.

Organic methods are fine, but I am also realistic about how long organics take to break down to become available to the plant. The microorganisms must first turn the organic product into various forms of decomposition in order for the nutrients to be available to the plant. This is one of the reasons I use a conventional vegetable growing mix. There are many good ones on the market. This mix is formulated with the proper nutrients and pH, (next section is pH) and will give you a really great start. Once you have used this for a season, you can easily amend it with your compost. How green is that!

Use bagged soil…don't ever buy from a pile. You never know where it came from, what has been sprayed in it, or how long it's been sitting there, exposed to the elements. Bagged soil is more reliable. My favs: Miracle-Gro and Pro-Mix BX.

The easy way to determine how much soil you will need to fill a container is to use an online soil calculator: www.gardeners.com/how-to/soil-calculator/7558.html. Or just simply multiply the length, width, and height of the container. Be sure that your measurements are in feet. This will give you the volume in cubic feet. Take this value and divide by the number of cubic feet of soil per bag of soil mix. This information will be on the bag. That answer will give you the number of bags you will need.

Don't forget your red wigglers! These super-composter worms can be bought at any fishing store.

Want to make the very best soil and not spend a penny? Wood chips from your friendly neighborhood arborist!! Many arborists will either give you the chips, which they sometimes have to pay to dispose of, or charge you only a minimal fee. In South Florida, you can go to www.nextdoor.com and sign up for wood chips, or visit chipdrop.com.

Spread the wood chips instead of piling them and they will decompose faster. Add them to your composter or you can even use them as mulch. After they have completely broken down, add them to your garden by sprinkling your composted wood chips on top of your mulch. As you water, the nutrients will go down into the soil.

A Little Bit About Soil pH

Soil pH is a measurement of its acidity or alkalinity. The number on the pH scale will give you that information.

Most vegetables prefer their soil to be in the acidic 5.8 -7 range. Higher than 7 is an alkaline soil.

If you want a reliable pH test on your soil, you need to take it to a good lab to get the correct results. The University of Florida offers reasonably priced soil testing through its Extension Soil Test Laboratories.
Their webpage is http://soilslab.ifas.ufl.edu.

Adding organic matter and nutrients to your soil naturally corrects pH. If you are starting with a reliable vegetable soil mix and using your own compost to amend the soil, the pH will be fine, and you can eliminate that step of having to get a soil test.

Do not rely on a $10 pH meter. Throw it out...it will not give you the correct pH.

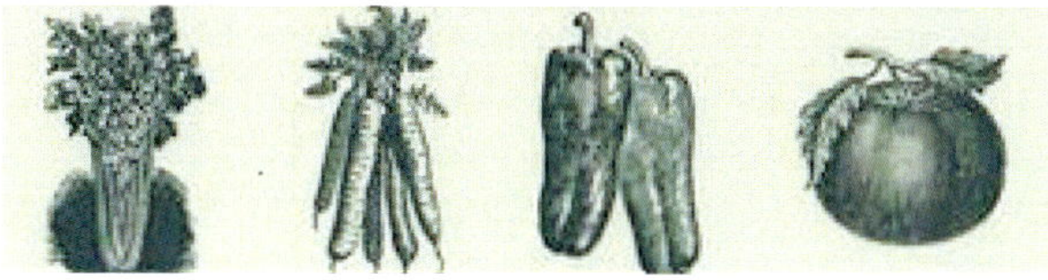

6. MULCHING

Yes, mulching is very necessary here so don't skip this. It holds water in and reduces weed growth. Mulch about 1-2 inches, so there are no empty spaces. Use straw, not hay, unless of course you'd like to grow it. Use the arborist chips, or pine bark nuggets if you want to be fancy!

Place the mulch around your plants, being careful to allow a space between the stem and the mulch, which can be an optimum spot for rot. Never mulch next to a stem. This is especially important when mulching fruit trees. The mulch should be at least four inches from the trunk to prevent rot, and to let oxygen in.

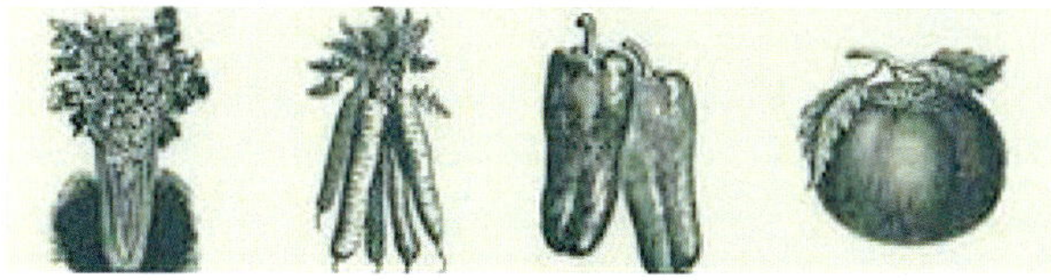

7. WATERING

How to tell when to water? Test several different spots around the plant and the garden by sticking your finger into the soil up to the second knuckle. If the soil feels damp at your fingertip, watering is not necessary. If the soil feels dry at any certain spot you want to give it a little water at that spot.

The best way to water a grow bag is to start at the outside and slowly work your way around to the middle. In a perfect world, you would water your plants from your own well. Municipal water contains chloramines, which can damage healthy bacteria in your soil. Chloramines cannot be taken out of the water, unless you use reverse osmosis, but you can reduce the chloride by setting your water out in a bucket for 24 hours.

As a general rule, keep the leaves as dry as possible. That doesn't mean when it rains you go outside and dry them off, it just means that wet leaves are more susceptible to foliar (leaf) diseases. This is one of the reasons it's best to water first thing in the morning.

If your plants droop, they may be telling you they're thirsty or too hot. There is no magic number of days for watering…it just depends upon the particular plant and the weather conditions. Just keep an eye on your plants, and check them daily using the finger test.

Some people invest in automatic watering systems. I think they are worth the extra cost. While there is no scientific evidence for this, based on my own observations, when plants know they are getting water every day, they don't get stressed. I get it. It's not like they can take a walk to get a glass of water. Plants see in colors and they know when it's a certain color its time to eat! Us too!

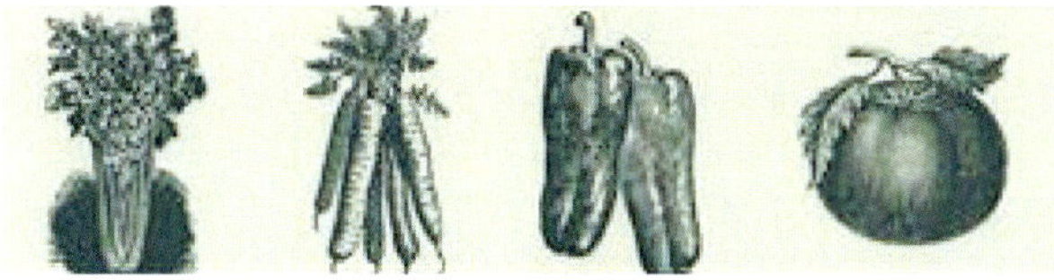

8. FERTILIZING

I usually advise a soil test before fertilizing, but because we are starting with a fresh, vegetable soil mix, this step is unnecessary. Start to amend the soil with your compost or fertilizer after your first season of growing. Sprinkle about an inch of finished compost on top of the garden each month to fertilize.

The tropics can put stressors on plants, so we need to feed our plants with whatever they might need to reach their full potential, so that they're strong enough to ward off pests and diseases.

Vitamin-rich vegetables require many different minerals and nutrients. I fertilize with either compost or fertilizer at the beginning of every month so I don't forget.

Not everyone has access to great compost so in that case, I would advise fertilizing each month at half strength using a fertilizer specifically formulated for vegetables. Fertilizer, at its molecular level, has the same exact elements that plants use and need.

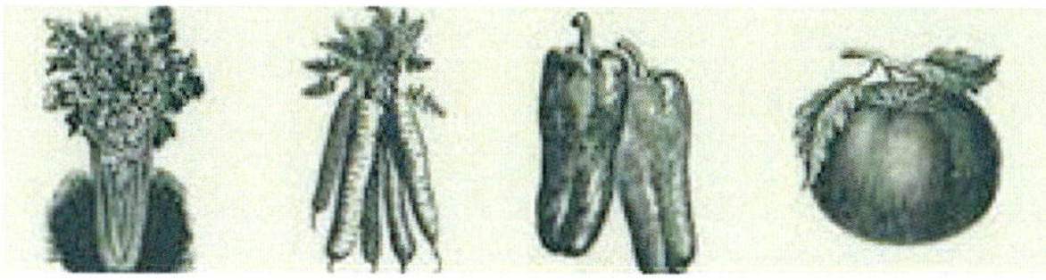

9. POLLINATING

To insure pollination, you can do it yourself. When you see the flowers, get a cotton swab or a small cosmetic brush. Start "Beein" and touch every open flower on that plant. Use a fresh cotton swab or small brush for each crop to avoid cross pollination. Do this in the morning when the pollen is fresh. Tomatoes and greens are self pollinating... no Beein necessary! Sometimes you may gently shake a tomato cage to assist in the process.

You can attract bees by planting an edible flower called "Borage." It's a cute little blue star-shaped flower that tastes like cucumber.

Plant flowers like marigolds in clumps to attract the pollinators to your garden. They will be more attracted to a clump, than a single flower.

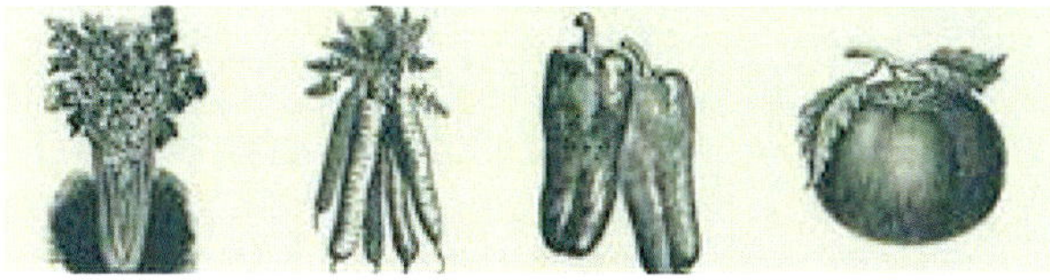

10. EASY-TO-GROW VEGETABLES IN THE TROPICS

Arugula is packed with nutrients and makes a wonderful addition to salads. Plant it in the fall, and harvest it by pinching off the outside leaves. Don't wait too long…fresh arugula is stronger tasting than what you get in the grocery store, so harvest when the leaves are small. Shade tolerant. Shallow roots. 6 per square foot or 8 in a bag. 35-40 days.

Basil comes in many different varieties and on most, we are experiencing a fungal problem. The leaves get bulls eyes on them. The cinnamon basil, or small leafed types, seem to be the more fungal resistant.

Pinch off the small basil leaves at the top so that the plant remains short and stocky. Remove the edible flowers when they appear or else the basil flavor will go into flower-making instead of the leaves. When harvesting basil, cut it just above a node or leaf juncture. Basil loves water. Shade tolerant. Shallow roots 4 per square foot or 6 in a bag. 75 days.

Beans: Kentucky Wonder pole beans are a great starter crop for beginners. Watch out for the bean leafroller (see page 91). Beans love to climb, so put in a trellis at planting time. They are self-pollinating. Shade tolerant. Medium roots. 9 per square foot or 11 in a bag. 55-85 days.

Beets are a root crop and prefer to be direct-sown, meaning sow the seeds directly into your bag or your bed. When you see the tops coming out of the soil, poke around to see if the beet is just about the size it would be in the grocery store. If you like the edible green tops, try the Chioggias. Their tops are prized by gourmet chefs all over the world. Shade tolerant. Medium roots. 16 per square foot or 18 in a bag. 45-65 days.

Borage is shaped like a star and is a small, edible flower that tastes a little like cucumber and attracts bees. You may have seen it…they use it to decorate ice cream. You can interplant borage with other crops to improve pollination. 55 days.

Broccoli: When it reaches about the same size as in the grocery store, cut the center head at a 45° angle to prevent water from pooling onto the cut stem. In a few weeks, baby broccolis will grow around the perimeter of the plant. Broccoli leaves are edible too, and are used in stir-fry dishes, but don't harvest them until the plant has finished producing the broccoli florets. Shallow roots. 1 per square foot or 1 in a bag. 85-100 days.

Cabbage worms pose a major threat to cabbage, so I stopped growing it. I'm still on the fence about it. If you want to grow it, regularly check under the leaves and brush off any eggs or insects. Shallow roots 1 per square foot or 1 in a bag. 90-110 days.

Callaloo is a type of bitter green from Jamaica that grows wild in the tropics. You can plant it in a bed, or in a bag in full sun. It can be grown in the ground as well, but it will be more susceptible to pests. Sauté this green…it doesn't steam well. Shallow roots. 4 per square foot or 6 in a bag. 60-70 days.

Carrots are a root crop and should be direct sown. Make sure your soil is nice and light with no rocks because when carrots hit something hard they will split or fork. To prevent this, you may want to mix perlite, which is puffed mica, into the soil to lighten it up to insure your carrots will be straight. Perlite can be purchased from any nursery.

When you see the carrot tops forming at the top of the soil, feel around them to see if they're big enough for your liking.

If you have small people living with you, get your cameras ready!!! Watching a kid pull a carrot out of the ground is hysterical!!! Carrots are ready usually after 65-85 days. Coreless varieties don't have those hard centers. Nantes are a great variety of coreless carrots. Medium roots, 16 carrots per square foot or 18 in a bag. 80-90 days.

Cauliflower: One plant will produce one cauliflower. Cauliflower can get tough under the hot sun. Look for a variety that says self-blanching on the packet. This variety will protect its head by having some leaves grow over and around it. If you don't get a "self blanching," when you see the head beginning to form, loosely tie the leaves over it, to protect it from browning in the sun. Harvest 75-90 days after planting. Shallow roots. 1 per square foot or 1 in a bag. 75-85 days.

Cilantro should be planted late November because it is sensitive to the extreme heat. If it gets too hot, cilantro will get leggy, flower, begin to produce seed, and taste bitter. You may want to try culantro. Shallow roots. 4 per square foot or 6 in a bag. 50-60 days.

Collards: Georgia Southern are very tasty! Look out for worms and snails. Best eaten when young. Shallow roots 4 per square foot or 6 in a bag. 70-90 days.

Dill can get tall and leggy. It loses its flavor in cooking so be sure to add it last. Deep roots. 8 per square foot or 10 in a bag. 60-70 days.

Eggplant loves the heat and may be grown in the summer months as well. The long Japanese varieties grow better in the tropics. Watch out for the little worms. Harvest by cutting the fruit from the plant. Use a cage for support. Medium roots. 1 per square foot or 1 in a bag. 90-110 days.

Garlic/Shallots: It can take up to 8 months to produce a bulb, and that's why I grow garlic, onions, and shallots for their chives. Take a bulb of garlic and separate the cloves, leaving the skin on. Plant the cloves one inch down in the soil with the pointy ends up. The chives come up fast, and are sweet and delicious. Shallots can be grown the same way. Wrap your chives in a damp paper towel and put them in the refrigerator to make them last longer.

Ginger: Plant fresh ginger 1 inch down in part sun into your soil with the knobby roots facing up. Water sparingly until the plant comes up to avoid rotting. Ginger takes ten months to produce, and can get up to four feet tall and three feet wide. It goes through a dormant period in the cooler months after the foliage dies back. It does not want or need to be watered until it comes back around May. 2 or 3 per square foot or 3-4 roots in a bag.

Herbs: All the herbs will grow very well down here year-round, with the exception of cilantro, which can have problems with the heat. All herbs are shade tolerant and prefer to be in dappled sun, maybe under a tree.

Kale: grows very well in the tropics. The small, young leaves are more tender than the bigger ones. As with all greens, pinch the outside leaves from the bottom at the core. There are several different types of kale, so you may want to try different varieties to see which ones you like best. Shade tolerant. Shallow roots. 4 per square foot or 6 in a bag. 70-80 days.

Leeks like full sun. Harvest when the stems get to be about one inch in diameter. Shallow roots. 12 leeks per square foot or 15 leeks in a bag. 60-100 days, depending on the variety.

Lemon Balm: I grow this just to smell it. Put this in your tea…no lemon required! Shallow roots. 9 per square foot or 11 per bag.

Lemon Grass gets very tall and wide and doesn't mind the heat. Don't plant this in a bed. You'll want to keep it in its container so it doesn't take over your growing area, or it can easily be grown in the ground as well! Pretty in your landscape because it makes a big pouf and edible too…it's a two-fer!

Lettuce is a fast growing crop and is ready for harvest in a couple of weeks. We grow leafy lettuce here and not heads. If the weather gets too hot, the lettuce will grow unusually tall and leggy, put out a flower, and start producing seeds. This is called "bolting." The lettuce, at this point, is very bitter and nearly impossible to eat.

Pick lettuce, or any leafy green from the outside, because it will continue to grow from the middle. Pinch it off from that bottom core and it will gift you with more lettuce. Shallow roots. 6 per square foot or 8 in a bag.

Marigolds: The bees are flying around and looking for a nice place to hang…coax them with a beautiful clump of marigolds! They use the flowers as landing platforms. Would you rather land on a twin or a king? There's your logic!

Mint seeds are tiny and difficult to germinate. It's easier to start from a plant. It's in the basil family so pinch the tops, just like you would with your basil, to maintain a small and bushy plant. Mint spreads everywhere like lemon grass. You can plant it in a bed but may want to plant it in a container so it doesn't take over the area. Shallow roots, 1 per square foot or 1 in a bag. There are many different types of mint available: chocolate, lemon, mojito, the list goes on and on.

Nasturtiums are edible, colorful, cascading flowers and are a great addition to any garden or salad. Asian restaurants sometimes put these in their salads. They have a peppery flavor. 1-2 per square foot or 1-4 in a bag.

Okra is a big plant, loves the hot weather, and prefers to be grown in the summer, even in the tropics. Cut the okra off the plant when it reaches 12 inches or else it will get too woody to eat. Deep rooted. 4 per square foot or 6 in a bag.

Onions: In the tropics, we plant green onions instead of bulbs. Harvest green onions after 50-75 days. 25 per square foot or 28 in a bag. You can also take your old onion cores/ bottoms and plant them to get chives!

Oregano: There are many different varieties of oregano. It cascades beautifully down the side of any container, making it a decorative addition to any garden. Because oregano is hardy and spreads as it grows, you don't need more than one plant. Shallow roots. 1 per square foot or 1 in a bag.

Parsley: Be patient with the seeds, they can take up to three weeks to germinate. There are two kinds of parsley – flat leaf and curly. A flat leaf variety called "Big Italy" makes a huge, beautiful parsley plant. Parsley will lose its flavor after the second year. Deep roots. 4 per square foot, or 6 per bag.

Parsnips grow just like carrots…Deep roots. 12 parsnips per square inch or 14 in a bag.

Peppers Sweet/Hot: The first bell peppers will be small and smashed looking until the plant gets large enough to support the weight of the bells, so pinch off those first few flowers. The plant's energy will then go into making the plant stronger to support the weight of the bells. For that reason, you may want to get a large sturdy looking starter plant instead of starting the bells from seed. Medium roots 60-100 days, depending upon the variety and size 1 per square foot or 1 in a bag.

Pineapples can take up two years to produce fruit. The green ones in the store are not ripe yet, and do not ripen off the vine. Once it's picked, that's its sweetness. If you have never experienced a ripe pineapple you are really in for a treat!

To get a real pineapple in all of its glory, it needs to be picked when it's golden yellow.

Find the biggest pineapple you can, cut off the top at the core, and tear off the first two to four rows of leaves to expose the brown dots, which is where the roots will grow from. Put the pineapple top in a glass, with it just touching the water. Use toothpicks to hold it in place so you don't drown it. Change the water every couple of days.

The white roots will grow down into the water. After the roots are at least an inch long, it's ready to plant right in the ground. You can also grow them in a raised bed or a container.

Plant the top where it will be exposed to full sun. Allow plenty of room because the leaf span can get up to 6 feet.

To plant: Dig a SHALLOW depression in the soil to accommodate the size of the top, and hill the soil around it to keep it upright and stable. Give it a little water in the center of its crown, and then never give it water again!

Pineapples are very self sufficient! The rain takes care of them and they are able to store the water they need.

The pineapple is ready when it turns from green to golden yellow, and you easily smell its sweet aroma of deliciousness. Harvest by twisting – not pulling – the pineapple off the plant like you would open a jar.

The plant will produce a new pup, but the fruit will decrease in size with each new generation. You can avoid this, and get a bigger pineapple, if you detach it from the mother plant and plant it by itself. Doesn't seem logical... who doesn't need their mommy!!

Pineapple plants in the landscape are a beautiful thing… another "two-fer!" I have over 100 plants and love to share them, and watch people's faces as they try one for the first time! Kids gobble them up! Here is another perfect example of involving children with free magic!

Radishes are very fast and very easy to grow. They are ready for harvest in only 20-30 days. The tops will poke up at the top of the soil. Try the Japanese varieties too… they grow like carrots. Shade tolerant. Shallow roots. 16 per square foot, or 18 in a bag.

Rosemary grows like wildfire down here. It can be planted in the ground or in a container. It's difficult to germinate the seeds because they are so tiny, so start from a plant. Medium roots. 1 per square foot or 1 in a bag.

Sage is not hardy in Zone 10b but can be grown here. Don't give your sage too much water, and let the soil dry out a little bit between waterings. To harvest, cut it at the top just above a leaf node, or juncture, to keep it bushy and short. Medium roots. 4 per square foot, or 6 per bag.

Seminole Pumpkin: This yummy, string-less pumpkin, cultivated by the Seminole Indians, tastes just like acorn squash. The vines can run for longer than 20 feet and love to climb. Plant them so they will grow up a trellis or a tree, but if you allow them to run along the ground, you need to protect them from nematodes.

Worms love to make their homes in these really soft leaves, so always check the leaves. Squish Deep roots. 1 per square foot, or 1 per bag.

Spinach is heat sensitive. Start your seeds late November. Harvest after 45-60 days.

Pick from the outside and always right before eating because spinach loses its nutrients very quickly after picking. It has shallow roots and is shade tolerant. 9 per square foot or 12 in a bag. "Bloomsdale Long Standing" makes an excellent spinach salad!

Squash/Cucumbers: I do not recommend these crops because they are very susceptible to foliar diseases. If you must grow them, I would suggest you grow them away from your garden, maybe on the other side of the house, so as not to bring pests or disease to your garden. Medium to deep rooted. 1 per square foot, or 1 in a bag.

Sugar Snap Peas: These are delicious but are not a heavy producer. You will need to plant a lot of them to get a large harvest. Use a cage for support. Harvest after 70 days. Shade tolerant. Medium roots. 8 per square foot, or 10 in a bag.

Sweet Potatoes can be grown here in the summer as well as in the fall. Sweets are grown from sweet potato slips, which are grown from the sweet potato. You can buy sweet potato slips directly from seed companies, which is only a one-time expense, but worth it because you will always have the mother plant, and can make slips for your family and friends!

You can sometimes make your own slips from sweet potatoes from a store if it has not been treated to retard budding.

To produce your own slips, place a sweet potato in a glass of water, keeping half the potato submerged and the other half above the water line. Stick toothpicks into the potato to keep it in place. In a few weeks, shoots will start to form at the top, and the roots will grow down into the water.

When the shoots have several leaves and are about three inches long, gently pinch them off at the base. Once you pinch them off, you can follow one of two procedures:

1. Place the bottom of the slip (the part that was detached from the potato) in about two inches of water and it will grow roots. When the roots are at least one inch long, plant the slips to just cover the roots with soil, leaving the leaves exposed on top of the soil.

2. You can also plant the slips directly into your container, but when you do it this way, you must you keep the slips moist, to get the roots growing. Keep them moist for at least a week.

Sweet potato leaves will run all over the place but the potatoes will remain where you planted the slips originally.

Feel around in the soil. As soon as you decide they are large enough to eat, the potatoes can be harvested one by one by pinching them off the vine. They don't have to be harvested all at once. The rest can remain in the soil for a while, and will continue to grow. 2 slips per square foot or 4 in a bag. 120-150 days. I like the Beauregards.

After harvesting sweet potatoes, you must CURE them in a cool, dark place for at least two weeks to sweeten them up. They store well and will last many months.

Swiss Chard has a lighter, buttery texture than other greens. After four weeks, pick the leaves from the outside of the plant first, so it will continue to grow from the middle, and you will be able to harvest continually until the plant dies out. Shallow roots. 6 per square foot or 8 in a bag.

Tomatoes come in two types:
Determinate
Indeterminate

The determinate varieties are compact plants, and the fruit will ripen all at once. Indeterminate tomatoes can get up to 10 feet tall, and the fruit will ripen whenever the fruit wants to! Tomatoes are very susceptible to foliar disease, so choose disease-resistant varieties by looking at the info on the seed packet. Always water at the soil level, being careful not to splash up and hit the undersides of the leaves.

To me, the smaller tomatoes taste great and my favorite is the Everglades tomato. It is native, and smaller than a cherry tomato, like a tomato berry. They are very prolific, mature quickly, and are deliciously sweet and very easy to grow in the tropics and South Florida.

Tomato suckering? Look at the photo (left). See the suckers? In the tropics, it's better to remove them so you will get more air circulating around the entire plant. You can use the suckers to make another tomato plant.

Our tomato lady, Marsha Eisenberg, who is featured in the next section, will share with us her secrets for growing large tomatoes in the tropics.

Turnips are a root crop and so should be direct sown. You can eat the tops, too! They're ready for harvest in 60 days. 8 per square foot or 10 in a bag.

There are other tropical crops but they may be more difficult, or maybe I haven't tried to grow them yet. (Next book!)

Summer Crops

- Callaloo
- Black Eyed Peas
- Hot Peppers
- Eggplant
- Okra
- Pineapples
- Asian Long Beans or Yard Long Beans
- Sweet Potatoes
- Cowpeas
- Malabar spinach
- Herbs (no Cilantro, but you might try Culantro, which is similar in flavor.

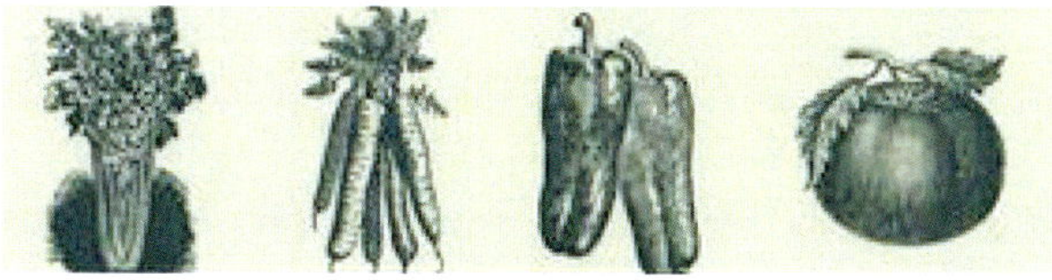

11. TOMATOES WITH MARSHA EISENBERG

Marsha Eisenberg is considered a tomato expert and her passion for tomatoes has led to a great deal of study, growing, and experimenting. Her techniques are specially adapted to raising tomatoes successfully, from seed to seed, here in South Florida, where the growing challenges are different from anywhere else. A speaker at local garden clubs, she is an avid seed trader with private breeders all over the world, and has many rare and unusual varieties

Marsha started to pursue her love of gardening – specifically, tomato gardening – after practicing dentistry for 22 years. She has generously agreed to share her secrets with us.

Here's her recipe for tomato success:

- Planter size: Minimum 18"diameter and 18" high (24 inches is even better).
- Full sun: Minimum 6 hours.
- Growing medium: peat based potting mix (Home Depot sells Lambert's). The mix should be fluffy and

not mucky – don't use any product with the word "soil" in the description! Her favorite is PRO-MIX BX. You can get this from Universal Enterprises Supply in Pompano Beach.

- Fertilizer: Any of these vegetable fertilizers: Vegetable Tone, Tomato Tone, Jobes Heirloom Tomato Food, Tomato Maker. 1 cup for every 5 gallons of mix.
- Soil amendment: lime-dolomite (to adjust the pH of the soil and to help prevent blossom end rot). 1 cup for every 15 gallons of mix.
- Pest control: In a sprayer, mix 1/2 tablespoon of copper fungicide, with a little less than a gallon of water and mix thoroughly. Add a tablespoon of Bacillus thuringiensis (Bt). Shake well and then add a few drops of dish soap or insecticidal soap. Coat the entire plant including the undersides of the leaves. Spray every 7-10 days. After a heavy rain, the next morning is an optimum time for preventative spraying. Spray early morning or late evening, because the sun can, and will, burn your leaves.
- Plant deep up to the first set of leaves. This allows for all the "hairs" on the stem to become roots and makes the plant stronger.
- Watering: tomatoes like moist-to-the-touch conditions but never standing water. Water the soil (base) and try not to wet the leaves.
- As the plant grows, it will need structural support –

trellising, cages, stakes, netting, etc.

- Once the plant starts to flower, fertilize every two weeks. Add 2 tablespoons per 5-gallon container.
- Prune all leaves up 12 inches from the soil line, so that no leaf touches the soil to avoid soil-borne diseases. Prune some of the internal leaves as it grows bushier for airflow, to keep fungus down.
- Always wipe the pruner with alcohol or a Lysol wipe before going to the next plant so you don't transfer a disease. Also, prune off all diseased leaves that have spots and yellowing. Bag them, and remove them from the garden. It's all about cutting disease transmission.

NEMATODES are the number one reason why tomatoes go from thriving to suddenly failing. They are microscopic worms that invade the tomato roots and cause root knots that prevent the plant from taking in nutrients. A simple preventive measure is to use fresh, new potting mix, and if you're using a grow bag or a pot, put the bag the potting MIX came in right under your pot. It acts as barrier to keep the nematodes from migrating into your grow bag or pot through the drain holes.

You may email Marsha at: Ginger2778@gmail.com

Marsha's Famous Tomatoes:

Elgin Pink

The Karma Breeding Project

Marsha's Tomato Explosion!

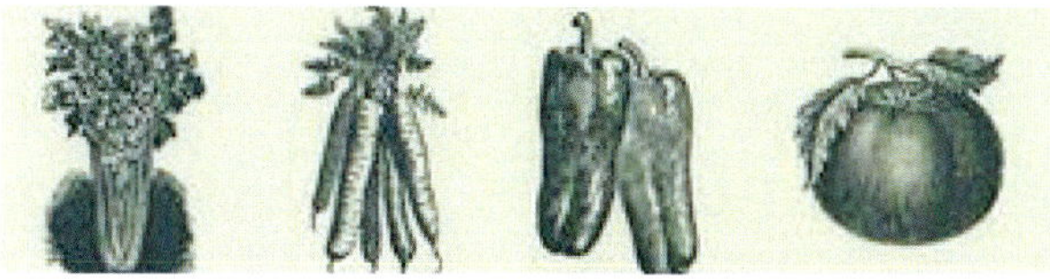

12. THE HARVEST

Start checking your vegetables wwhen they are just about the same size as they are in the grocery store.

The optimum time for harvesting is first thing in the morning when your food is at its highest peak of nutrition and flavor. Store it in the refrigerator, or pick it right before you're going to eat it.

Some things like peas, taste better when they're young. Pick it off and try it! Picking encourages fruiting, so pick away. Pinch it off the plant or cut it off close to the stem.

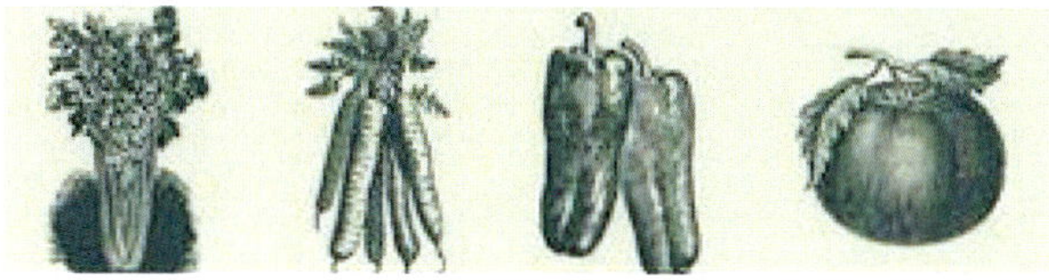

13. SHADE TOLERANT PLANTS

These are plants that can tolerate a couple of hours of shade each day and still be happy. Keep in mind that we are growing in the winter, which has less sun anyway, so you're bounty may be smaller.

To keep it simple and as a general rule…keep your veggies in mostly sun, and keep your herbs in mostly shade.

These shade tolerant plants are:

- arugula
- basil
- beans
- beets
- collards
- herbs
- kale
- lettuce
- mustard greens
- peas
- radishes
- spinach
- Swiss chard

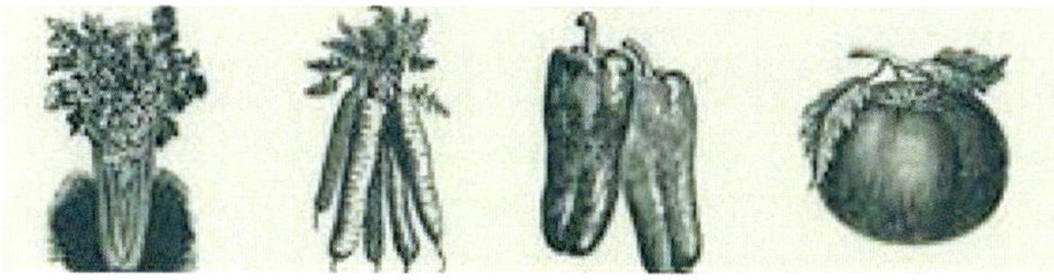

14. EDIBLE TREES

When you grow an edible fruit tree, it's best to get a grafted tree from a reputable nursery or fruit tree expert. The fruit will be ready in 1-3 years. Get a dwarf, unless you like climbing trees. Buying a dwarf fruit tree will save you the expense of investing in a monkey.

As a general rule: With fruit trees, it is better to have a couple near each other to improve pollination and get more bounty.

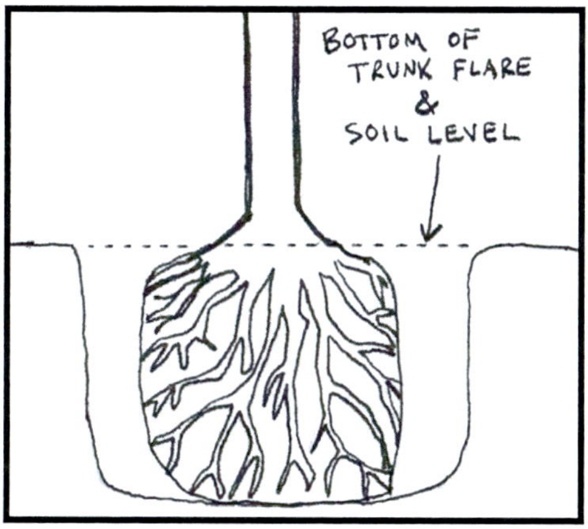

Always plant a fruit tree, or any tree, at its flare. The flare should remain above the soil line. Refer to page 37.

To plant a fruit tree or any tree in the ground:

- Loosen the tree by gently rolling the container.
- Carefully remove the tree by its stem and inspect the roots.
- If they are circling, tear them apart or cut them. We want its roots to go out into the soil like the spokes on a wheel.
- Shake off the excess potting material and save it in the container.
- Dig a hole in the ground twice the diameter of the plant roots and no deeper than its root ball.
- Place the tree in the hole, making sure the root flare is above the soil line (see picture above).
- Fill in with our native soil or with whatever soil you have in that area, and stabilize it.
- Soak it with water.
- Mulch at least four inches away from the trunk.
- Use the material you saved in the container and put it on top of the soil.

I am currently growing: Avocado, Banana, Lime, Lemon, Mango, Moringa, Miracle Fruit, Cranberry Hibiscus, and Pomegranate.

Fertilize your citrus trees with a fruit tree fertilizer. Just follow the package directions. Less is better than more.

Because of the citrus diseases we have had in South Florida over the last twenty years, our fruit has visible damage on the rind, but it won't affect the fruit on the inside. It's fine to eat…I don't know of a human who has caught citrus canker or citrus greening. The university and many scientists are working on this and hopefully will have a solution soon.

Avocado: Florida has a native avocado called the "alligator pear" but it's a little watery, and not as good as the Hass avocado from California. There is a cultivar called "Florida Hass" and it's pretty good!

Bananas are native and tropical so plant them right in the ground. Start from a small tree. Bananas love water. They will be ready for harvest about a year after planting. The first ones will be ready for harvesting approximately 12 weeks after the first flower opens. You can write that date directly on the banana flower. Pick them when they are plump, green, and almost ready to turn yellow. Ripen the bananas in the shade or in a cool, dark place. Enjoy your bananas now because bananas have a disease and we may lose them. Dwarf Cavendish is delish!

After the harvest, the tree will fall over and die. Let the old tree stay there...or cut it up. It will nourish the new ones. A new tree will begin to grow near the previous one.

Cranberry Hibiscus: The leaves of this tree taste just like cranberries. Put the leaves in a salad to give it a nice zing! It grows like weed here and reseeds itself everywhere.

Mangos: The Haden, Edward, and Nam Doc Mai are all string-less and delish! There are so many varieties and everyone has their own favorite.

Miracle Fruit Tree produces a small, red, yummy berry. When you put it in your mouth, it tricks your taste buds into thinking that anything sour has magically turned sweet. Doctors who know about this berry give it to their cancer patients who, as a result of chemotherapy, have a metallic taste in their mouths and stop eating. This berry gets food tasting good again. Give this to a picky kid who won't eat healthy fruit...they will eat it now because it tastes better!

THIS IS A MIRACLE! Lemons taste like sweet lemonade! Oranges taste like nectar from the Gods!

Moringa Trees are native to Africa and parts of Asia, and are one of the world's most nutrient-dense foods. Every part of this tree is edible. Plant this tree right in the ground and it will get very big, very fast! It doesn't need much water or fertilizer, and grows very lush. The leaves are tasteless...put them in your salad...tell your kids it's a new kind of lettuce...they'll never know they're eating healthy!

Papaya Plant: Plant this tree in full sun or else the sweetness of the fruit will be compromised.

There are many other tropical fruit trees that grow here, however, I haven't tried to grow all the varieties, so I cannot tell you about them.

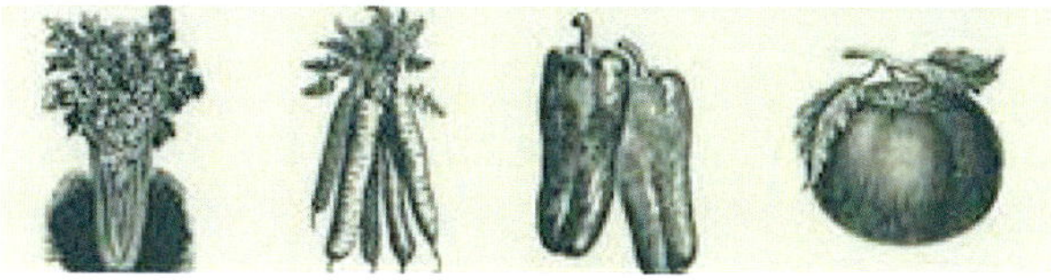

15. COMPOSTING

One of the best and cheapest fertilizers around is compost that you make yourself. Rich in nutrients, compost makes an ideal fertilizer for your plants. Being able to produce your own fertilizer contributes to sustainable gardening and is the the greenest way to go.

Choose a spot for your composter that provides mostly sun…it can go under a tree and have some shade too in the tropics. Sometimes it can heat up so much you can lose microbes, but they will come back eventually…it's just nature.

Buying an inexpensive composting barrel from a garden supply store is the easiest way to start composting. I use the one that has a trap door at the bottom, so it's easy access to the oldest compost first. Some fancier models have handles that allow you to rotate the barrel and mix the compost. You don't need something elaborate…mixing it will make it turn faster, but nature will always turn it. Always cover your composter.

To compost, place anything that was once alive in your composter, except meat, fish, and dairy. Use your yard waste and lawn clippings if they haven't been sprayed with anything. That spray could dissipate in the compost process but better safe than sorry. If a plant looks diseased or compromised, don't put it in your compost pile. Err on the side of caution. Use your food waste.

One of the best natural fertilizers is seaweed. Go to the beach at dusk when the seaweed has washed up on shore. Put it right in your composter. Ocean water is full of nutrients so add a little ocean water to the compost too. Use a cup or two…not too much because of the salts. We are doing this to increase the microorganism diversity in our compost. We want as much diversity as possible because we don't know enough about the soil-food web to have a specific answer as to decomposition, and how long it takes.

Put leaves over your fruit waste or it will attract a lot of fruit flies and start to stink…and then your neighbors will not like you. I wouldn't either!

Compost is done when it's light and fluffy and has an earthy odor.

Red Wigglers are the super composter worms. Put them right in your garden and/or in your composter... they will be happy there. Investing in a vermicomposter is unnecessary.

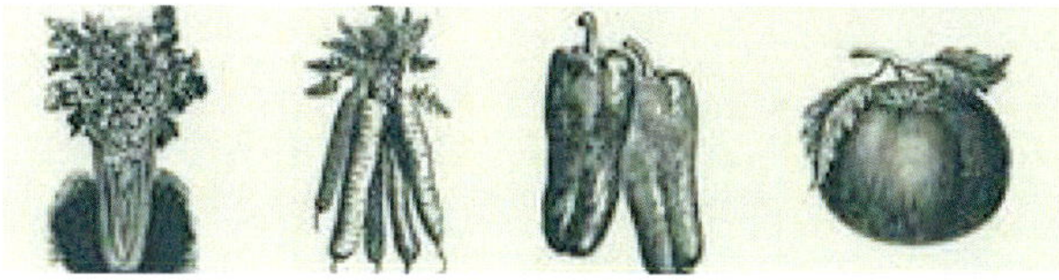

16. PLANTING WITH KIDS - MY FAVE!

I haven't met a kid who wasn't fascinated and delighted by the magic show that goes on in the planting world. Engage your children. Talk about "Eating the rainbow!" For little kids, have a contest to see who can eat the most colors! Plant pineapple tops.

Grow carrots. Carrots come in a variety of colors. Kids are so amazed when they see the carrot come out of the ground and they look shocked when they see this magic! It's so much fun to watch them!

Have an ugly carrot contest by not lightening up the carrot soil to see what sizes and shapes can happen!

Lacewing

Praying Mantis

Ladybug

Lacewings, ladybugs or praying mantises are beneficial insects that pray upon some of the bad guys, but don't make the mistake I did and buy them to release in the garden for the kids.

They will fly away as mine did, which made the kids cry.

Please learn from my mistakes and don't make the kids cry!

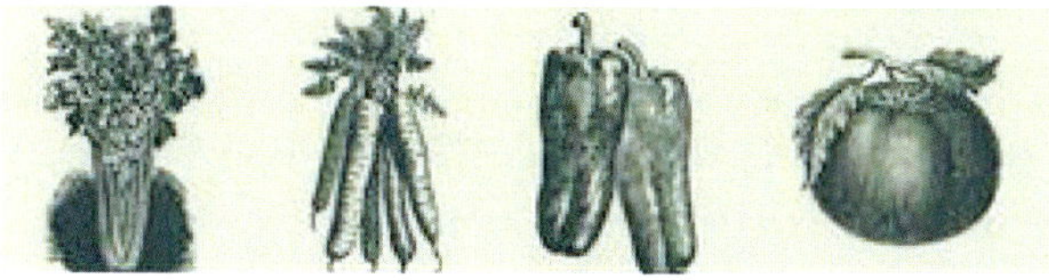

17. GROWING FOOD FROM FOOD

Onions: Every time you use one of these, make sure you save the bottom. Cut it 1/2-inch up from the bottom. Place it in the soil, cut side up, and cover lightly. In a few weeks, you will have onion chives that are delicious and sweet, unlike those expensive ones in the grocery store.

Garlic: Plant a single clove with the skin on, one inch down, pointy side up and you will get garlic chives.

Cabbage and Lettuce planting can be done from a core. From the core, cut 1/2-inch up. Root it in water, with it just touching the water line. Use toothpicks if you need to hold it in place. After two weeks you will see the roots, and it should have some leaves on the top, too. Plant it at the level where the top growth appears.

Ginger/Turmeric: Plant one inch down, knobby sides up in dappled sun…under a tree perhaps. It will take 10 months to produce. Ginger and turmeric will grow dormant during the winter, and come back to life around June. Do the same thing with a turmeric root.

Peas and Beans: Remove them from the pod, dry them out, and plant them in soil about one inch down.

Pineapples: This one's explained in the Easy-To-Grow Crops list on page 53.

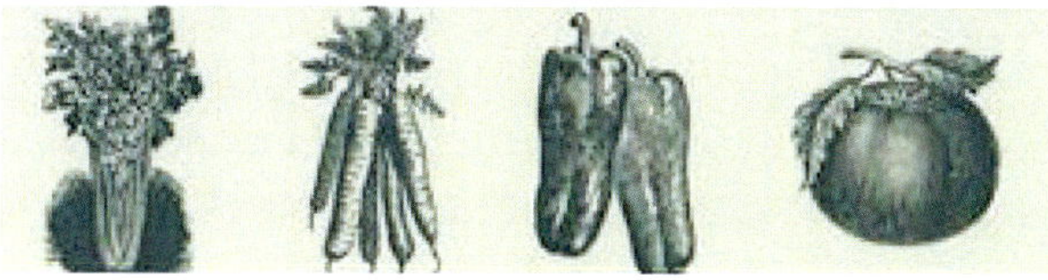

18. SPROUTS IN A JAR

In general, any plant from which you might eat the stems and leaves is a good option for sprouting. Commonly used seeds include mung beans, lentils, daikon radish, broccoli, red clover, barley, and wheat. However, there are many others. Items you‘ll need:

- A jar with a wide mouth
- Seeds – buy "sprouting seeds"
- Straine or cheesecloth

There are several sprouting systems on the market. They provide the most convenient way of growing sprouts. They are also expensive. Just use a glass jar and either a strainer or a piece of cheesecloth fastened across the opening with a rubber band.

Put two tablespoons of seeds in a jar and cover them with two inches of water. Do not cover it with the lid…keep it open. Let it sit on the counter overnight. In the morning, place a strainer over the mouth of the jar, or cover it with cheesecloth, and drain the water.

Pour out the soaking water and rinse with fresh water.

Drain the rinse water. The seeds should be slightly damp. At this point, you are not soaking the seeds as they will rot. Rinse twice per day until the sprouts are the size you want. The seeds should be covered with something to keep out dust and insects. Store your excess sprouts in a covered glass jar in the refrigerator.

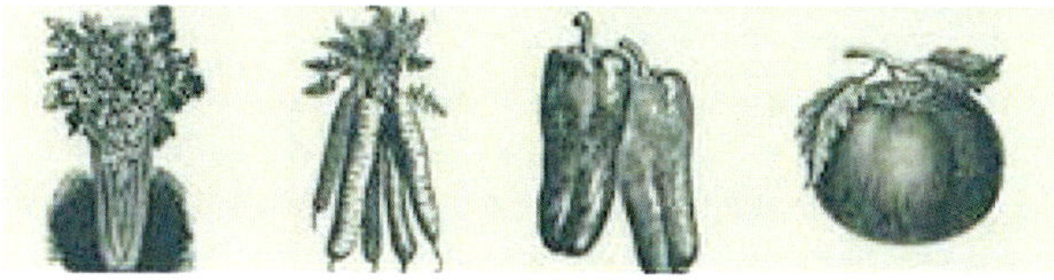

19. EAT THE WEEDS

DANDELIONS: If you are picking these from a yard, make sure it hasn't been sprayed with pesticides. Harvest the small leaves and the flowers. Sauté the greens with onions and garlic. Coat the flowers in egg and flour. Deep fry.

JAPANESE KNOTWEED tastes like a cross between rhubarb and a tart apple. Harvest the shoots when they are no longer than 8 inches or they will be too tough. Use it in any recipe that calls for rhubarb.

KUDZU looks very similar to poison ivy, and can be found almost anywhere. The leaves, flowers, and potatolike roots are all edible. Use the leaves like spinach. Kudzu produces beautiful flowers that have a grape aroma, and can be made into a jam or syrup.

LAMBS QUARTERS, also known as wild spinach, is higher in nutritional content. The seeds, related to quinoa, are very edible. Sauté this green with lemon, garlic, shallots, and olive oil. You're using a shallot, so now you're a gourmet weed eater!

PURSLANE is everywhere. Eat it raw or sautéed, and still young. It contains a huge amount of Omega3s and vitamins. Epicurious.com has several recipes.

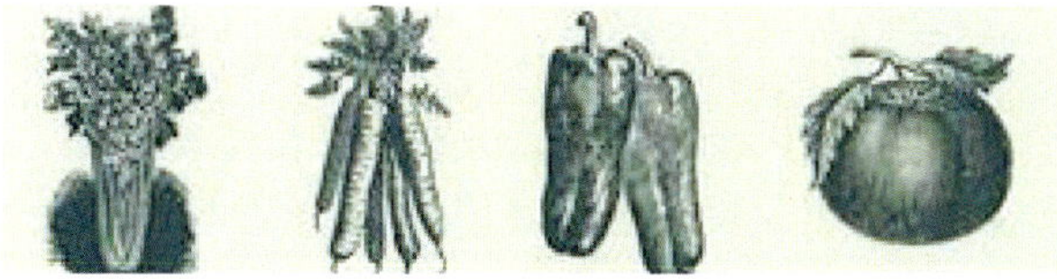

20. PESTS

Aphids: When you start to see leaves curling, sooty looking mold, and a lot of ants running up and down your plants, you might have aphids. It may be too much nitrogen in the soil. It's easy to get rid of them. Just a sharp spray of water from the nozzle of the hose will usually do the trick. Do this in the morning so that your leaves will dry off during the day. If your infestation is severe, you may have to spray at dusk as well.

Bean Leafrollers love to make their homes in the bean leaves. They are voracious eaters and will destroy the leaves on your plants. It's easy to tell they are there because they cover themselves with a tiny leaf blanket. Squish the blanket, then you don't have to touch them.

Cabbage Loopers and Worms move like inch worms and are usually found on your broccoli, cauliflower, and cabbage. Pick them off.

Cucumber Beetles can be found on many crops, not only cucumbers. These beetles can be striped or spotted. Pick them off.

Cutworms hide under the soil during the day and come out at night to eat the stems of your plants, making the plant fall over and die. You can put a paper roll around the seedling to protect it, or try making a collar out of tin foil and placing it around the base of the plant you think might be in harm's way. I've never had this problem.

Flea Beetles are tiny black bugs that make pockmarks in the leaves. Use your yellow sticky traps, explained in the next section, to lure them away from your edibles.

Grubs are the larvae of various beetles. They are the fat white worms that you see in the soil…use tweezers to pick them out.

Iguanas can decimate your garden in no time. The only thing that can keep them away is a custom-made iguana cage made from pet screen that is very strong. We've built a cage out of pet screen and it is working!

Leafminers cause squiggly white lines in your leaves. This is made from the larva from an insect, which eats the leaf tissue from the inside. As soon as you notice this happening on your leaves, take the leaf off and throw it in a sealed plastic bag in a garbage can.

Mealybugs are strange, cottony looking bugs that attach themselves to our plants in cottony puffs because they like to hang together. They can spread disease. Use your soap and a toothbrush to knock them off. Drown them in a bowl of soapy water. Wash the plant completely after any application of soap.

Scale looks like bumps. Those bumps will suck the life out of your plants. Try using a soft toothbrush with soap and scrub gently.

Snails and Slugs feed on a variety of living plants and on decaying plant matter. Slugs chew irregular holes with smooth edges in the leaves. Snails chew small round holes in the leaf. To eliminate snails or slugs, just pick them off…or if you're squeamish use tweezers.

Spider Mites are barely visible to the human eye. They make silk web like structures on the leaves to protect them from predators. There are many different species. Wash them off with a sharp spray of water.

Stinkbug is called the "shield bug" because of its shield-like appearance, and its ability to emit a foul odor when it feels threatened. Pick them off and hold your nose!

Sri Lanka Weevil is a whitish gray pest that wears a shield. This guy bites notches out of the edges of the leaves. It's about the size of a ladybug. The best thing to do is to go out in the morning or at dusk and drown them in a bowl of soapy water.

THRIPS are hard to see without your Nerd Glasses (see next chapter). If your leaves are distorted, you may have these. A sharp spray of water will cure that one.

TOMATO HORNWORMS are caterpillars that can rapidly defoliate your plants. You can distinguish these from other caterpillars by the protruding horn on their backside. Pick it off and drown it in a bowl of soapy water. Use tweezers if you don't want to touch it.

Whitefly is serious. We have several different varieties in South Florida. When you see eggs under the leaves, brush them off. The fly is small and very quick…they're hard to squish. However, when you have so many whiteflies that it looks like it's snowing, try using Sevin or hire a professional to spray with Carbaryl.

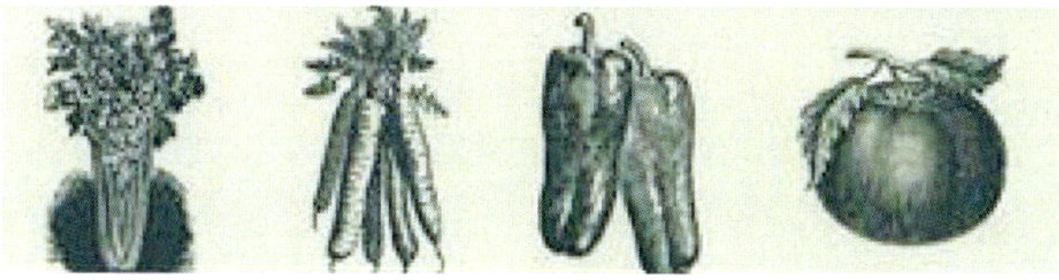

21. PEST CONTROL

Start by choosing the right types of plants for Zone 10b because these plants are better adapted to our extreme weather. As a general rule, pests go for the weaker plants.

Spraying destroys the healthy bacteria and fungi on the leaves, making plants more vulnerable to insects and diseases. Farmers have to spray. Gardeners don't have acres of plants, so we can simply watch our gardens and pick these pests off. I spray only when I have run out of options.

The most important way to control pests naturally is to prevent infestation before it becomes a problem. You can prevent an infestation if you look at your leaves daily, and check the undersides where pests love to hang.

This is how I deal with pest problems: When the sun goes down, I put on my Nerd Glasses and look under the leaves. I pick, squish, brush off eggs, or drown what I find in a bowl of soapy water.

In the mornings, so the leaves will not stay wet all day, I'll use a sharp spray of water to get the pests off. Use your hose nozzle as a gun. They may come back but if they do, they will be at a different spot and will have to start all over.

Shaklee makes a great natural soap called Basic H. I use it for everything, even in my home. A natural soap won't destroy the waxy protective coating on the plant. Use a couple of drops of Basic H (or any other natural soap) mixed with a gallon of water in a spray bottle. This makes the pests unhappy, and makes the plant slippery. Wash the entire plant after, to get rid of any residue.

If it is snowing whitefly, I will have to use a broad-spectrum insecticide called Sevin. The bottle screws right onto the end of the hose. It is so simple. I use this only as a last resort when an insect infestation threatens to destroy an entire garden. I haven't had to use this product for many years.

If you decide to spray anything on your plants, don't spray when the sun is out because it will burn your leaves.

Neem oil, made from the seeds of the Neem Tree. It's a strong broad-spectrum "organic" insecticide that can help to solve many pest problems and may prevent some foliar diseases. Follow the directions on the package. Mix in a drop of Basic H to help the Neem better adhere to the leaves. For some reason, organic people use this as their go-to product because it is from the Neem Tree. I never use it. Even the smell makes me gag. How would you like to be sprayed with Neem Oil? I can't imagine a plant would like it much either. Nope.

Bt, an organic solution for worms and caterpillars. Some people use it. I don't...I just take them off!

Diatomaceous earth, which is ground up fossils from the ocean. Organic gardeners like to use this for worms, but it could possibly harm your red wigglers too. I don't use it.

You may want to try using row covers. You can buy them, or use tulle, which is a netlike material found in a fabric store. Over your garden, make a frame or a tunnel-like structure with a sturdy thick wire so the material won't touch your plants. Place the fabric over the wire tunnel and secure it. Pile dirt on the edges of the tulle to prevent insects from crawling under the edges. Or another option would be to use landscape pins. The tulle will let the sun in and keep the pests out.

Use yellow sticky traps to see who is coming to visit. Pests are attracted to the color yellow, and get caught on the sticky substance on the yellow paper. You can buy these traps or make them at home. Get some heavy yellow paper and coat it with a sticky substance. You can use Tangle-Trap. Glue it to a stick and plant it in the soil by your plants. When the paper is full, simply throw it in the trash and replace it with another one.

Pest control powders aren't a good idea because powder will not stick to the undersides of the leaves. Use a liquid. Put some tin foil or sparkly stones around your plants. The sunlight will bounce up and confuse the bugs. Boil marbles...they shoot the light everywhere.

It's kind of a guessing game. We are always trying to figure out which is better...this or that. You have to weigh these challenges and decide which way is the best for you.

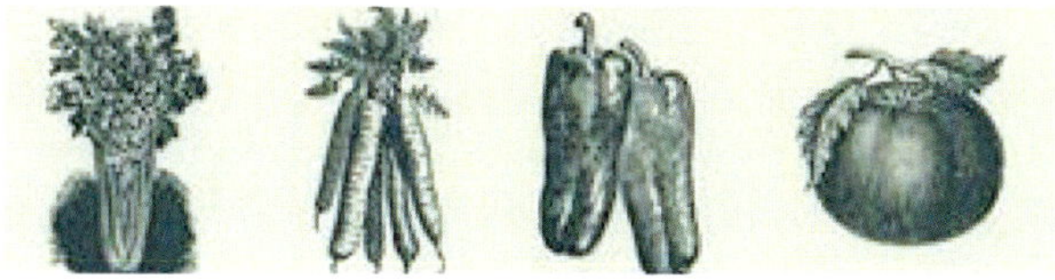

22. DISEASE

It is very difficult to diagnose a plant just from the damage. It needs to be looked at under a microscope by a plant pathologist to correctly diagnose the plant disease. If someone looks at your plant and tells you what disease you have, they may just be guessing, it may be an educated guess, but it is guessing.

Some of the symptoms of disease can include: white powdery spots, yellowing, brown spots that look like bulls-eyes, and black spots on the leaves and fruit.

At the first sign of any of these symptoms you should take action immediately, but usually by the time you notice them it's too late.

Sometimes you can stop these pathogens with a copper-based fungicide, but success is rare.

The best prevention is knowledge and good practices.

Use that mulch to create a barrier that will eliminate the splash up from the rain and watering

Take out all the diseased leaves. Repeatedly wipe your cutting tools with rubbing alcohol to disinfect them. Dispose of the cuttings in a tightly sealed bag so as not to spread the spores.

Blossom End Rot first appears as a round sunken area on the blossom end side of the fruit. The lesion gets larger and larger and will eventually rot out the fruit. The latest scientific evidence indicates that uneven watering creates a calcium deficiency. That's why I like a watering system. The plant is given the same amount of water every day.

Powdery Mildew can be fatal. If you have this I would be very careful… only water the plant when absolutely necessary. Use the finger test, and don't get any water on the leaves. Cut off the unhealthy leaves.

Rust is caused by a fungal parasite. Cut off the disease and hope it will heal itself.

Septoria Leaf Spot starts on the bottom of the leaves after fruit set. Plants seem to wither away and die from the bottom. Yellow surrounds the dark spots. Loss of foliage causes a decrease in the size of the fruit.

Sooty Mold happens when your plants are planted too close together, and the air isn't able to circulate around them. It is a thick black oily substance. You can wash it off with soapy water.

Southern Blight: Some of these symptoms include a white moldy substance on the stem, and a small water-soaked lesion on the lower stem where it meets the soil. The lesion will spread and cause the plant to wilt and fall over.

Verticillium Wilt is a fungus that lives in the soil. Part of the leaf turns yellow, then brown. Tomatoes often get yellow shoulders. There is nothing you can do about this. This is the one time where you will have to get rid of the tomato soil. Don't even compost it.

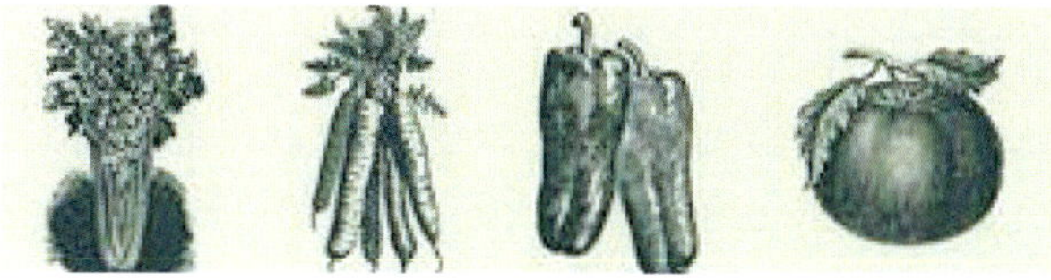

23. MISCELLANEOUS TIPS

- Don't touch your plants when they're wet. It can spread disease.
- Walking on your garden soil will compact it. Soil is alive. You could smush something.
- Do not crowd your plants.
- Water carefully at the soil level…overhead watering in the tropics is not a good practice.
- Compost tea does nothing…it's a myth and a pollutant. There are scientific comparative studies on compost tea and its efficacy.
- Eggshells do not decompose and do not add calcium to the soil. Sharp shells could harm your red wigglers too.
- Cardboard is a haven for termites, and inhibits the flow of oxygen and water through the soil.
- Epsom salt is for the bath, and Tums are for the stomach.
- Nothing is needed in your garden in your first season except your new vegetable gardening "Mix." It is

balanced correctly. The soil is filled with everything it needs to give you a great garden. Adding anything at this point is unnecessary, and could harm your soil's natural balance.

- After your first growing season, you will want to amend the soil because its nutrients have been somewhat depleted from its prior use. You can amend it by using a vegetable fertilizer once a month at half strength, or ultimately by using your own compost. Sprinkle about an inch of finished compost over the entire garden each month. Add fresh new soil to the top. These are good practices. I do this at the beginning of every month so I don't forget.
- Clean your tools with rubbing alcohol after using them so as not to spread disease…like a surgeon.
- Start small. Ten grow bags or one 4 X 8 garden bed per family is a good start.
- Store your seeds in a cool, dark place.
- Never touch a furry caterpillar – it stings
- There is so much we don't know about plants. Some of us, including me, are just better guessers!

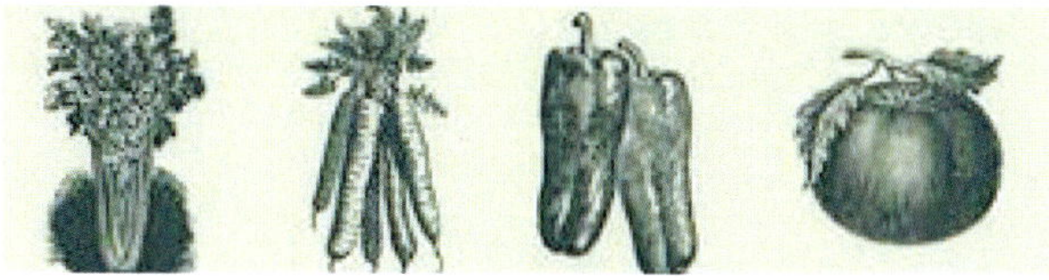

24. SUPPLIER LIST

Criswell Farm and Market, (954) 667-7611

Seeds:

Eden Organic, 954-382-8281
www.eonseed.com
www.JohnnySeeds.com

Supplies:

www.UFI.org; 954-696-9577

www.Gardenworkssupply.com; 954-537-1111

Pest Control:

www.Karencoulson.myshaklee.com
for Basic H

Fruit Trees:

Spykes Nursery in Davie

Miracle Fruit:

www.Miraclefruitfarm.com; 786-220-4135

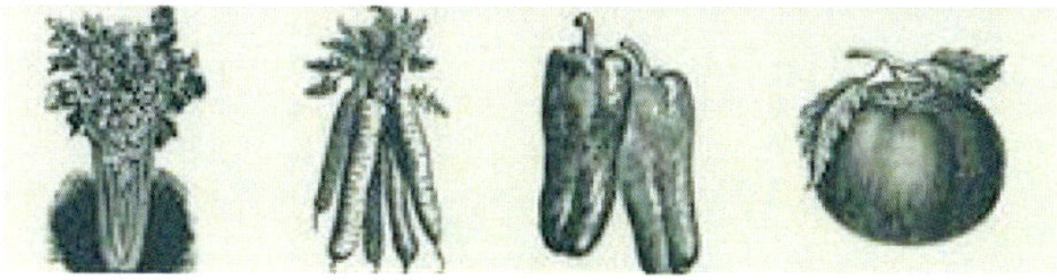

CONCLUSION

This book is written for the novice. Some of these tips you can slide on as you accumulate more knowledge. That's okay because my goal was to give you the best-case scenario for success, especially on your first try!

I hope I have streamlined this process, kept it simple, and ended any confusion you might have as to growing here.

Gardening is a great way to bond with your children or grandchildren. I cannot tell you how many times someone will come up to me and say… "I remember when me and my grandmother…"

Kids love to watch things grow, and through gardening, they learn valuable skills and lessons that will benefit them for the rest of their lives. Many of these "teachers of tomorrow" will be going into the new green businesses. Engage them early! We need great minds in agriculture.

And as for us, gardening is nearly free and always magical, creating beautiful memories and fun that lasts forever in our hearts and in our lives. Trust me on this one…

Happy Harvesting!

ABOUT THE AUTHOR

Karin Fields was born in Highland Park, Illinois. She currently resides in Fort Lauderdale, Florida, with her husband of over twenty-five years, her three cats, and her endless amount of edible plants.

Karin won 2 awards out of over 400 volunteers from the State of Florida for her work with children:

- Guardian ad Litem of the Year 2002, Miami
- Lorraine Munroe Spirit of Excellence Award

www.TheEdibleGardeningGal.com